A TRUE BOOK™

OUR UNIVERSE
THE
SUN

D1239844

Cody Crane

Children's Press®
An Imprint of Scholastic Inc.

Content Consultant
Hsiang Yi Karen Yang, PhD
Assistant Professor
Institute of Astronomy
National Tsing Hua University
Hsinchu, Taiwan R.O.C.
(Former Computational Astrophysicist and Assistant Research Scientist in the Department of
Astronomy at the University of Maryland, College Park)

Library of Congress Cataloging-in-Publication Data
Name: Crane, Cody, author.
Title: The sun / Cody Crane.
Other titles: True book.
Description: New York : Children's Press, an imprint of Scholastic Inc., 2020. | Series: A true book |
    Includes index. | Audience: Ages 8-10. | Audience: Grades 4-6. | Summary: "Book introduces the reader
    to the sun"— Provided by publisher.
Identifiers: LCCN 2020004519 | ISBN 9780531132210 (library binding) | ISBN 9780531132395 (paperback)
Subjects: LCSH: Sun—Juvenile literature.
Classification: LCC QB521.5 .C732 2020 | DDC 523.7--dc23
LC record available at https://lccn.loc.gov/2020004519

Design by Kathleen Petelinsek
Editorial development by Priyanka Lamichhane

All rights reserved. Published in 2021 by Children's Press, an imprint of Scholastic Inc.
Printed in Heshan, China   62

SCHOLASTIC, CHILDREN'S PRESS, A TRUE BOOK™, and associated logos are trademarks and/or
registered trademarks of Scholastic Inc.

Scholastic Inc., 557 Broadway, New York, NY 10012

1 2 3 4 5 6 7 8 9 10 R 30 29 28 27 26 25 24 23 22 21

**Front cover: The sun's fiery surface
gives off a golden glow.**

**Back cover: An illustration of the Parker
Solar Probe observing the sun.**

Find the Truth!

Everything you are about to read is true *except* for one of the sentences on this page.

Which one is **TRUE**?

T or F   The sun is made up mostly of a gas called helium.

T or F   The sun is the largest object in our solar system.

Find the answers in this book.

Contents

An illustration
of the sun's size
compared to the
eight planets in
our solar system.

The **BIG** Truth

An illustration of a red giant star

Will the Sun Ever Burn Out?

3 Observing the Sun

4 Big Discoveries

The Orion Nebula

Introduction

If you look up at the night sky, you will see many twinkles of light. Most of these are bright, burning stars, and our universe is filled with them. The sun is just one of billions of trillions of stars scattered throughout space. But for Earth, no other star is as important as the sun.

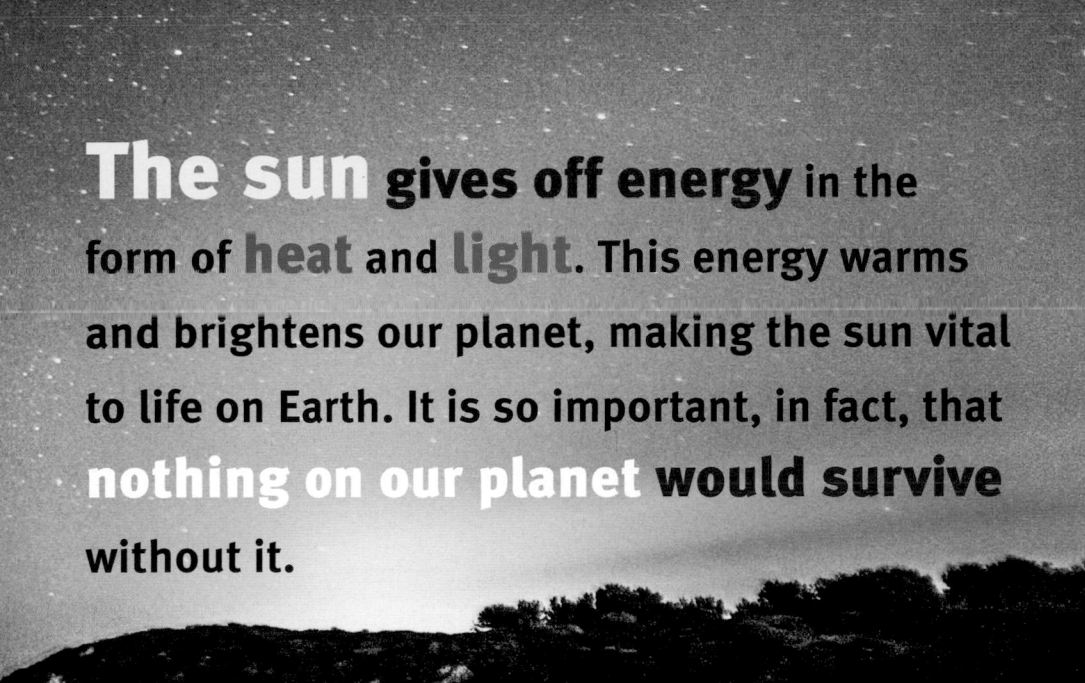

The sun gives off energy in the form of **heat** and **light**. This energy warms and brightens our planet, making the sun vital to life on Earth. It is so important, in fact, that **nothing on our planet** would survive without it.

The sun as seen from space.

The second closest star to Earth is Proxima Centauri. It is about 25 trillion miles (40 trillion km) away.

Our Star

The sun is a medium-size **star**. Like all stars, it is a huge, fiery ball in space made up of extremely hot gases. The sun is also the closest star to Earth. That is why it looks bigger and brighter than other stars we can see from our planet. Despite being the nearest star to Earth, it is still about 93 million miles (150 million kilometers) away. Imagine you could reach the sun by car. To drive there from Earth would take about 177 years!

Scientists have discovered some planets that orbit their star within the habitable zone. And they continue to discover more.

Earth might have looked like a land of snow and ice if the planet was much farther from the sun.

Sweet Spot

Even though the sun is really far away, it is the perfect distance from our planet. Earth lies in the sun's habitable zone. This is an area around a star where liquid water can exist, and possibly support life. If our planet was closer to the sun, Earth would be so hot its seas would boil away. If Earth was farther away from the sun, it would be a frozen wasteland.

A Planet Full of Life

Scientists believe about eight million different species of plants and animals live on Earth. Our planet is so far the only place in the universe known to support life. The sun's warmth and light make that possible. Plants need the sun's light to grow. Many animals rely on plants for food. Humans also rely on plants and animals to survive. Without the sun, there would be no life on Earth.

Unique animals and a variety of plants create colorful diversity on our planet.

Center Stage

The sun sits at the center of our **solar system**. Earth is one of eight planets that **orbit** the sun. The solar system is also home to moons that orbit the planets. Earth has one moon. Other planets such as Jupiter and Saturn have dozens. Smaller dwarf planets such as Ceres, Eris, and Pluto are part of the solar system, too. There are also icy comets and rocky asteroids.

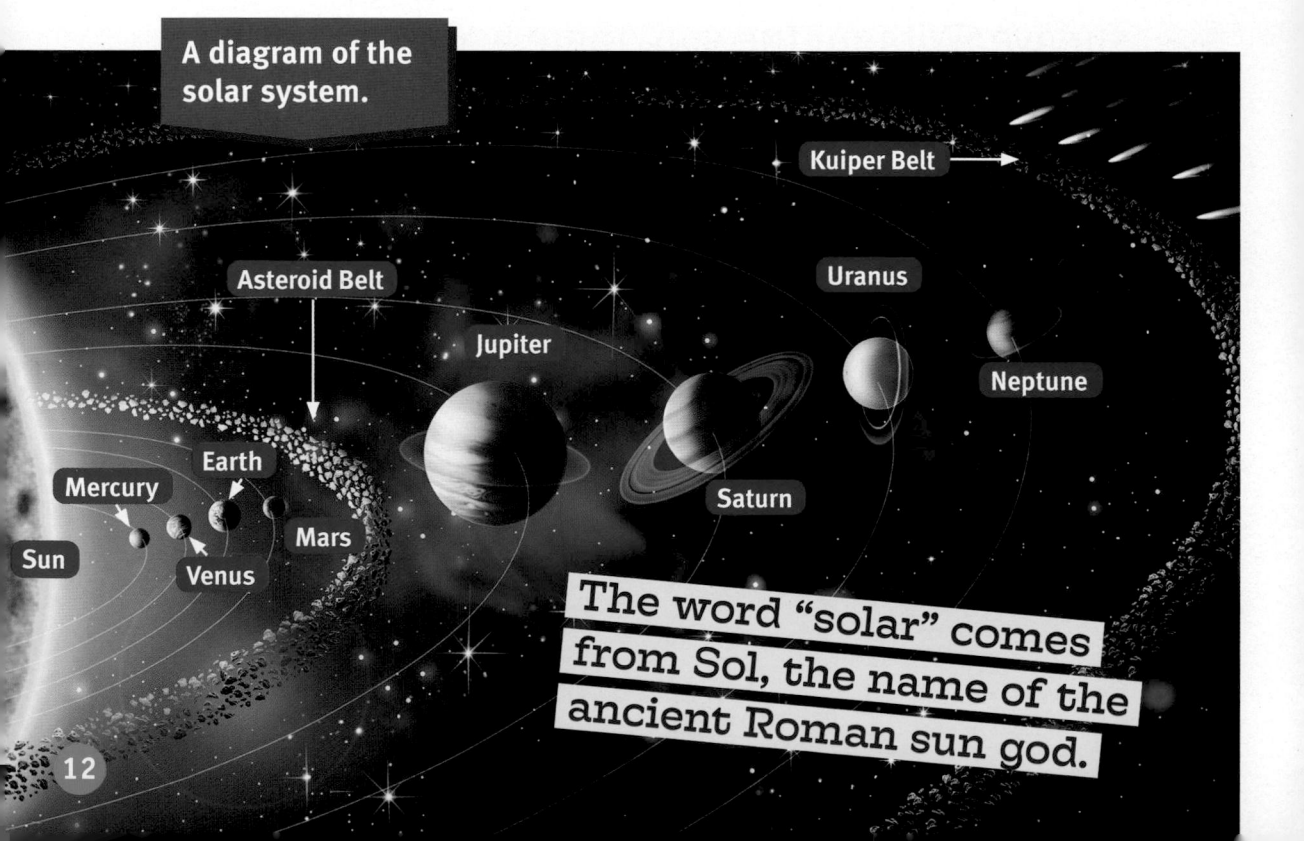

A diagram of the solar system.

Kuiper Belt

Asteroid Belt

Uranus

Jupiter

Neptune

Earth

Mercury

Saturn

Mars

Sun

Venus

The word "solar" comes from Sol, the name of the ancient Roman sun god.

The asteroid that may have killed the dinosaurs left a giant crater in what is now the Yucatán region of Mexico.

Sunblock

Did a lack of sun wipe out the dinosaurs? Many scientists think so. They believe a huge asteroid, or space rock, struck Earth about 66 million years ago. The asteroid was more than six miles (10 km) wide. The impact crushed rocks on the ground into dust and blasted the particles into the air. So much dust clouded the skies that it blocked out the sun. Without the sun's heat and light, temperatures dropped. Plants died and animals had nothing to eat. Three-fourths of life on Earth, including all the dinosaurs, died.

If the sun was the size of a beach ball, Earth would be the size of a pea. The other planets in our solar system are also much smaller than the sun.

Supersized

The sun is really BIG. More than one million Earths could fit inside it! The sun is not just large in size. It also has the most **mass** of any object in our solar system. The greater an object's mass, the greater its **gravity**. Gravity is a force that pulls objects together. The sun's gravity tugs everything in our solar system toward it. This is what keeps nearby objects in space, including planets, in orbit around the sun.

Circling the Sun

The length of each planet's orbit around the sun is different. It takes Earth about 365 days, or one year, to make a full trip around the sun. As planets orbit the sun, they also rotate, or spin. As Earth orbits the sun, it rotates on an **axis**. It takes 24 hours, or one day, for Earth to complete a rotation. As Earth spins, different sides of the planet face toward or away from the sun. This is what makes day and night.

Length of Days and Years for Planets in Our Solar System

PLANET		LENGTH OF DAY	LENGTH OF YEAR
MERCURY		4,223 HOURS	88 DAYS
VENUS		2,802 HOURS	225 DAYS
EARTH		24 HOURS	365 DAYS
MARS		25 HOURS	2 YEARS
JUPITER		10 HOURS	12 YEARS
SATURN		11 HOURS	29 YEARS
URANUS		17 HOURS	84 YEARS
NEPTUNE		16 HOURS	165 YEARS

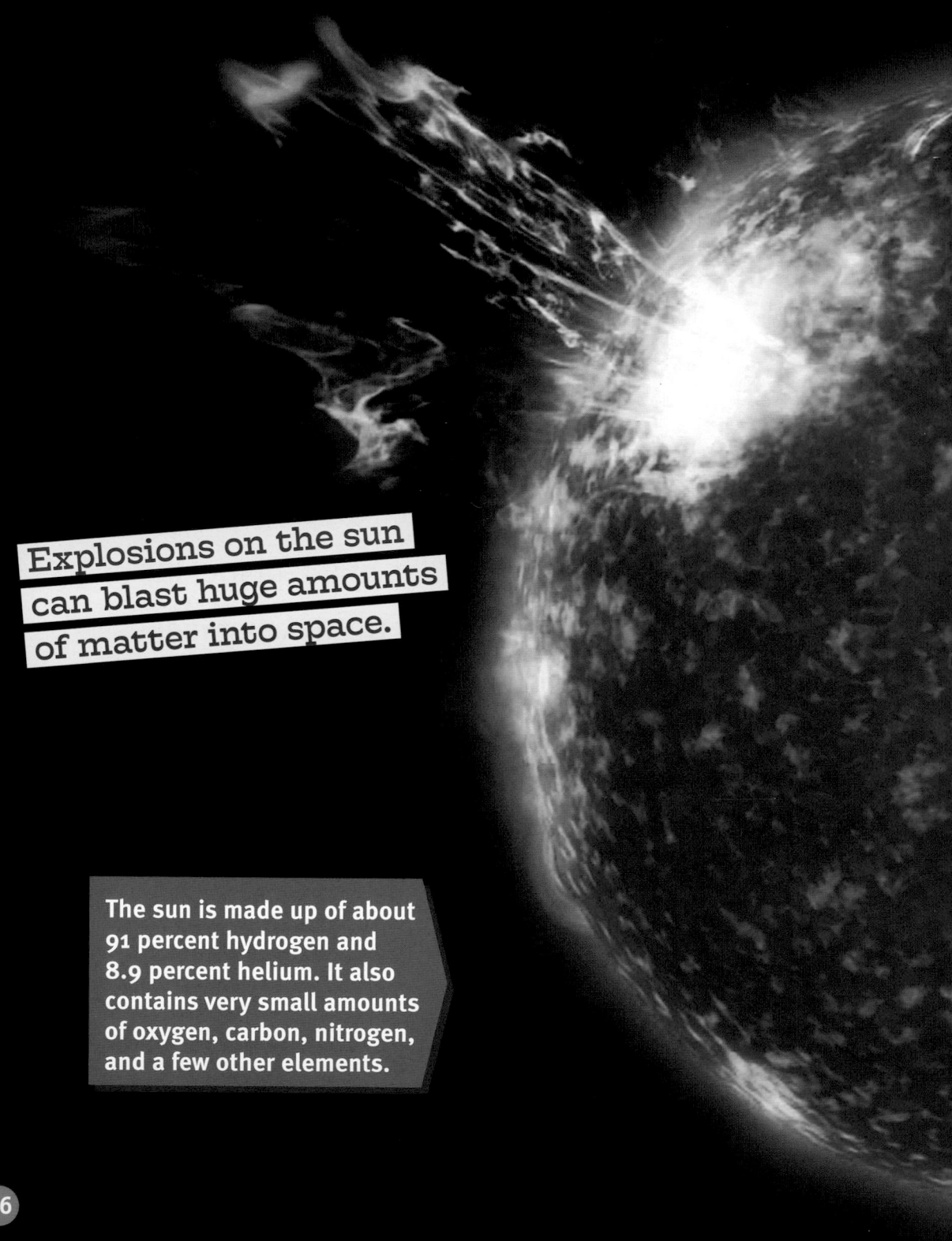

Explosions on the sun can blast huge amounts of matter into space.

The sun is made up of about 91 percent hydrogen and 8.9 percent helium. It also contains very small amounts of oxygen, carbon, nitrogen, and a few other elements.

Powerhouse

The sun is made up mostly of a gas called hydrogen. Hydrogen gets pushed and squeezed tightly into the sun's core, or center, by the strong force of the sun's gravity. The enormous pressure at the center of the star causes some hydrogen **atoms** to combine with one another. This process, called **fusion**, forms a new gas called helium. Fusion also produces a HUGE amount of energy. That energy keeps our fiery sun burning brightly.

Blazingly Hot

The temperature at the sun's core can reach 27,000,000 degrees Fahrenheit (15,000,000 degrees Celsius)! The surface of the sun is cooler. But it still reaches a scorching 10,000°F (5,500°C). The heat and light the sun produces spread out from its surface through space. Only one-billionth of the total energy the sun gives off reaches Earth. But that amount can still be intense. Luckily, Earth's atmosphere blocks some of the sun's rays. This protective layer of gases acts like sunscreen for the planet.

Children play in the water on a toasty 100°F (38°C) day.

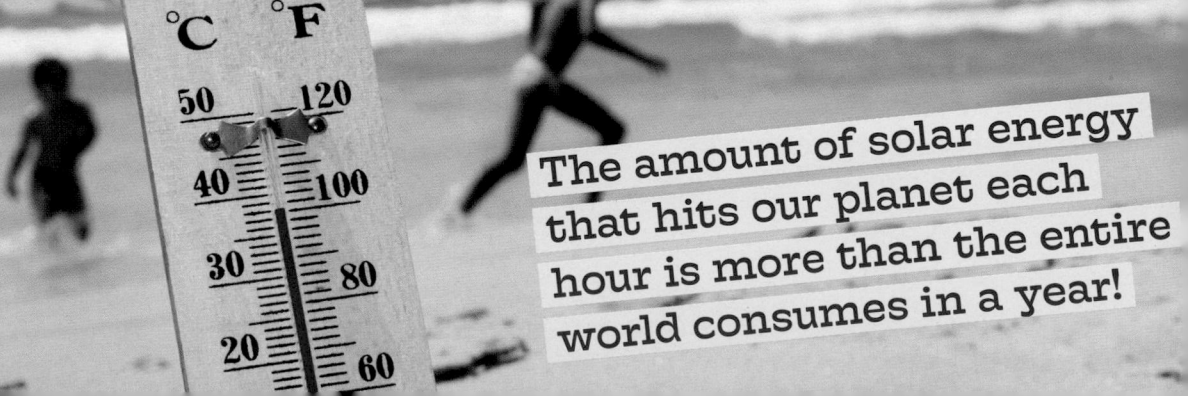

The amount of solar energy that hits our planet each hour is more than the entire world consumes in a year!

Parts of the Sun

The sun is made up of four main layers: the core, the radiative zone, the convection zone, and the photosphere. It also has a lower atmosphere known as the chromosphere, and an upper atmosphere, known as the corona.

It takes several hundred thousand years for energy to make its way from the sun's core to its radiative zone.

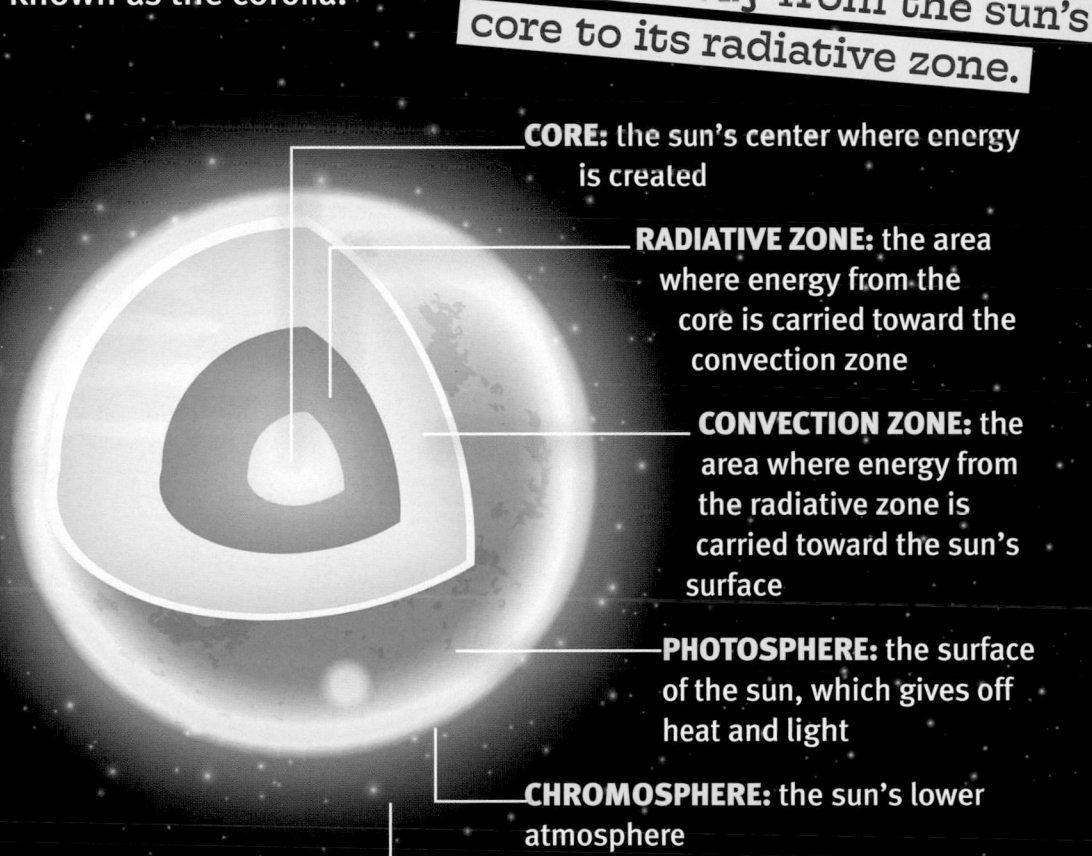

CORE: the sun's center where energy is created

RADIATIVE ZONE: the area where energy from the core is carried toward the convection zone

CONVECTION ZONE: the area where energy from the radiative zone is carried toward the sun's surface

PHOTOSPHERE: the surface of the sun, which gives off heat and light

CHROMOSPHERE: the sun's lower atmosphere

CORONA: the sun's upper atmosphere

A view of the Orion Nebula, a region of space where new stars are being born.

A Star Is Born

The sun is about 4.6 billion years old. Like all stars, it was born inside a **nebula**. This huge cloud of dust and gas in space is like a nursery for baby stars. It took millions of years for the sun to form. Gravity pulled bits of the nebula's dust and gas together. The material clumped into a spinning ball that continued to collect more and more dust and gas. The ball grew bigger and hotter until it burst to life as our sun.

An Average Star

Stars go through a life cycle and can change size throughout their lifetime. Our sun, right now, is a medium-size and medium-temperature star called a yellow dwarf. The largest stars in the universe are called giants and supergiants. Giants can be hundreds of times larger than the sun. Supergiants can be thousands of times larger. The smallest stars are called dwarfs. Along with different sizes, stars can also be different colors. Read on to find out why!

There are stars that are much bigger and much smaller than our sun.

Size Comparison

BLUE GIANT
(5 to 10 times larger than our sun)

RED DWARF
(the smallest stars)

YELLOW DWARF
(our sun)

RED SUPERGIANT
(hundreds to more than a thousand times larger than our sun)

Color Code

Our sun shines a brilliant yellow in the sky. Other stars are blue, orange, or even red. Color depends on the star's surface temperature. Red stars are the coolest. The next coolest stars are orange and then yellow (like our sun). The hottest stars glow blue.

A view of about 100,000 stars captured by the Hubble Space Telescope. Their colors shine brightly.

Red dwarfs are the most common types of star. They are also the smallest and dimmest.

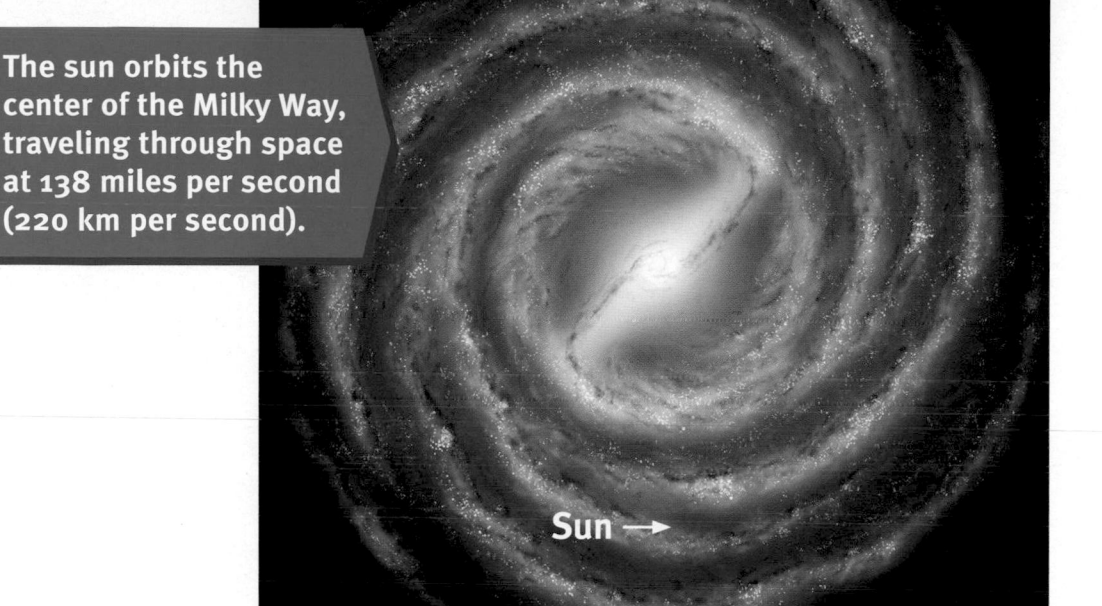

The sun orbits the center of the Milky Way, traveling through space at 138 miles per second (220 km per second).

Sun →

Not Alone

Our sun may be the only star in our solar system, but it's surrounded by many neighbors. It is clustered together with billions of other distant stars in space. These stars, which are at different stages of their life cycles, make up our galaxy, the Milky Way. The Milky Way is shaped like a giant pinwheel, with all of its stars, including our sun, spinning around its center. There are trillions of other galaxies in the universe, each containing billions of stars.

Will the Sun Ever Burn Out?

The sun is about halfway through its life. One day it will run out of the hydrogen fuel at its core and die. But don't worry. This won't happen for a very, very long time. It will take another five billion years to be exact.

The Life of the sun

Check out the phases of the life of the sun. As it goes through its life cycle, it will continue to change.

Birth

About 4.6 billion years ago, inside a nebula, gravity attracted dust and gas together. They formed a spinning disc. More dust and gas collected at the disc's center. It heated up. Our sun was born.

Grown-Up

Our sun is now a yellow dwarf. This type of medium-size star burns for billions of years.

Old Age

In five billion years, when the fuel at the sun's core runs out, its outer layer will expand about 100 times larger than it is now and shine 1,000 times brighter. At this stage the sun will be a red giant.

Last Gasp

Eventually, the star's core will collapse into a white dwarf. Its outer atmosphere will be flung into space, forming a ring-shaped planetary nebula.

The End

After several billion more years, the star's core will cool and stop glowing completely. This is called a black dwarf.

The Parker Solar Probe has a heat shield that can withstand temperatures up to 2,500°F (1,400°C).

The Parker Solar Probe was launched on August 12, 2018. It is on a nearly seven-year-long mission to study the sun.

Observing the Sun

The energy given off by the sun is powerful. People cannot get too close to it. They can't look directly at it without harming their eyes, either. So scientists have had to find other ways to study the sun. They view it using solar telescopes and other special instruments. Solar telescopes are specially created to view the sun without damaging eyesight. Scientists also send spacecraft to orbit the sun in order to learn more about it.

Scientists use solar telescopes during the day. Most other telescopes are used at night.

The McMath-Pierce Solar Telescope at Kitt Peak National Observatory was built in 1962.

On the Ground

To view the sun from Earth, scientists use ground-based solar telescopes. The world's largest solar telescope is found at Kitt Peak National Observatory in Arizona. Light from the sun enters the top of the telescope. It shines down a long arm that extends underground. The light bounces off mirrors inside the arm and into an observation room. There, scientists can safely view the sun's image on computer screens.

In Space

Although it's challenging to study the sun up close, scientists have launched several spacecraft to view the star from a safe distance in space. These solar probes orbit the sun and take images. They also collect data about what is happening inside the sun and on its surface. The Helios 1 and Helios 2 space probes were launched in the mid-1970s. In 2018, the Parker Solar Probe was launched. It soon flew closer to the sun than any other solar probe.

The Parker Solar Probe can fly fast enough to travel from New York City to Tokyo in less than a minute!

Distances of Solar Probes from the Sun Compared with Planet Mercury

Planet Mercury
36 million miles

Helios 1 Probe (1974)
29 million miles

Helios 2 Probe (1976)
27 million miles

Parker Solar Probe (2018)
4 million miles

A Different View

The sun looks yellow as it shines in the sky. But it actually gives off light of all colors and **wavelengths,** some of which are beyond the range of our vision. Some solar telescopes and space probes can detect these forms of light. These observations can reveal areas of different temperatures on the sun. This helps scientists learn about features on the sun's surface and changes in the energy it releases.

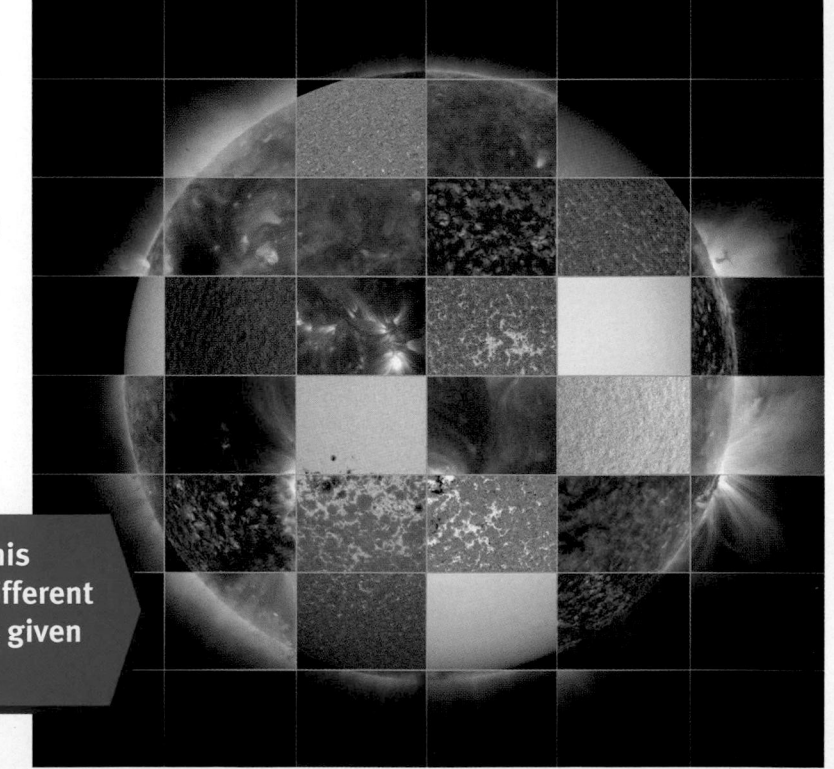

Each rectangle in this image captures a different wavelength of light given off by the sun.

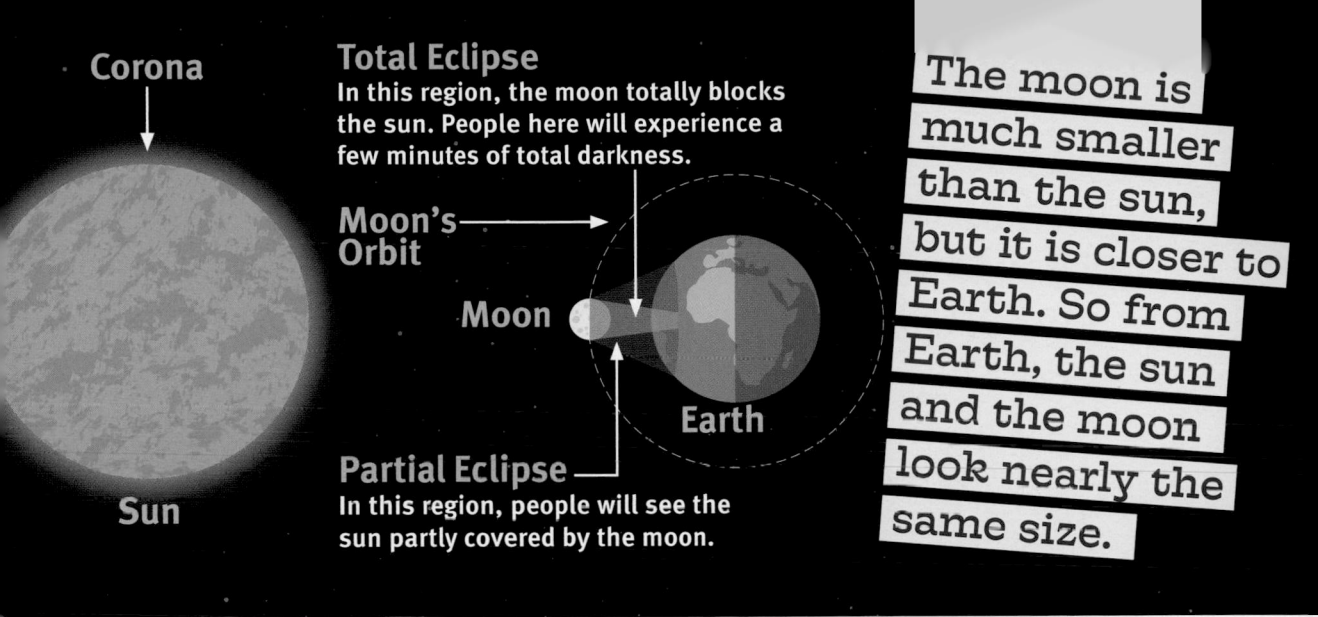

Corona

Total Eclipse
In this region, the moon totally blocks the sun. People here will experience a few minutes of total darkness.

Moon's Orbit

Moon

Earth

Partial Eclipse
In this region, people will see the sun partly covered by the moon.

Sun

The moon is much smaller than the sun, but it is closer to Earth. So from Earth, the sun and the moon look nearly the same size.

Solar Eclipse

About twice a year, the moon passes directly between Earth and the sun as it orbits the planet. This causes a phenomenon called a solar eclipse. During a solar eclipse, the moon blocks part or all of the sun's light from reaching Earth. For a brief time, this casts a shadow on Earth. During a total solar eclipse, when the moon completely covers the sun, scientists can observe the sun's corona. These hot, glowing gases surrounding the sun are usually hidden by its bright light.

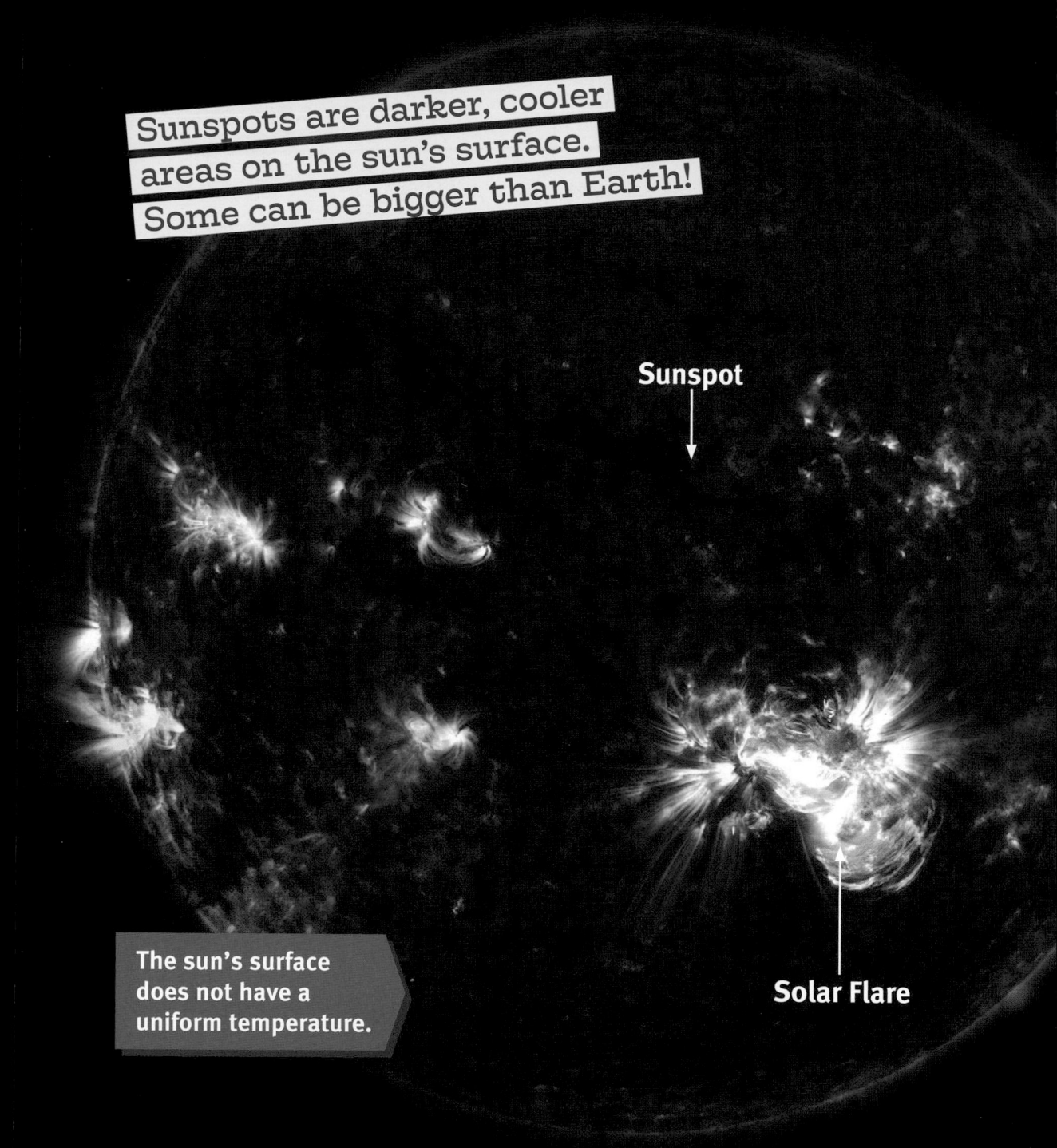

Sunspots are darker, cooler areas on the sun's surface. Some can be bigger than Earth!

Sunspot

The sun's surface does not have a uniform temperature.

Solar Flare

Big Discoveries

People have been studying the sun for thousands of years. And they still haven't unlocked all of its secrets. One thing scientists have discovered is that the sun is an extremely active place. Its surface is constantly bubbling, swirling, and changing. They have also discovered that the sun follows an 11-year cycle. During that time, its activity level rises and falls. You'll learn more about the sun's activity in this chapter.

Magnetic Field

Activity on the sun is caused by its strong magnetic field. This invisible field surrounds the sun and extends far beyond the edges of our solar system. You can picture the sun's magnetic field as many imaginary lines looping from its surface. These lines stretch, twist, and move over time. This happens as the sun spins on its axis, causing its gases to rotate at different speeds.

Timeline of Early Sun Discoveries

3340 BCE: Earliest record of a solar eclipse, found in Ireland.

1514: Nicolaus Copernicus claims the sun is the center of the solar system.

1609: Johannes Kepler describes how the planets orbit the sun.

Spots on the Sun

More than 400 years ago, scientists Galileo Galilei and Johannes Fabricius both separately observed dark patches on the sun. The dark patches were sunspots. These areas are cooler than the rest of the sun's surface. They usually form in pairs and last just days or weeks. Sunspots signal changes in the sun's magnetic field. More sunspots indicate more activity on the sun's surface. Fewer spots indicate less activity.

1611:
Galileo Galilei and Johannes Fabricius are among the first to observe sunspots on the sun.

1687:
Isaac Newton describes the idea of gravity.

1862:
Angelo Secchi determines the sun is a star.

1908:
George Ellery Hale establishes sunspots are areas with strong magnetic fields.

Solar Wind

The sun is constantly releasing into space a stream of energized particles known as solar wind. It flows in all directions from the sun at about one million miles per hour (1,600,000 km per hour). The solar wind collides with the protective magnetic field that surrounds Earth. The interaction can cause auroras to appear across the night sky. These colorful, glowing ribbons of light become larger and brighter when the sun is more active.

The solar wind causes auroras on Earth. This aurora glows over the country of Iceland.

How Auroras Form

1. The sun's solar wind blasts high-energy, fast-moving charged particles toward Earth.

2. Earth's magnetic field blocks the particles, redirecting them around the planet.

3. In regions where Earth's magnetic field is weaker, the particles penetrate deeper into Earth's atmosphere.

4. The particles collide with atoms of oxygen, nitrogen, and other gases in Earth's atmosphere. Each collision sparks a different colored burst of light, creating an aurora.

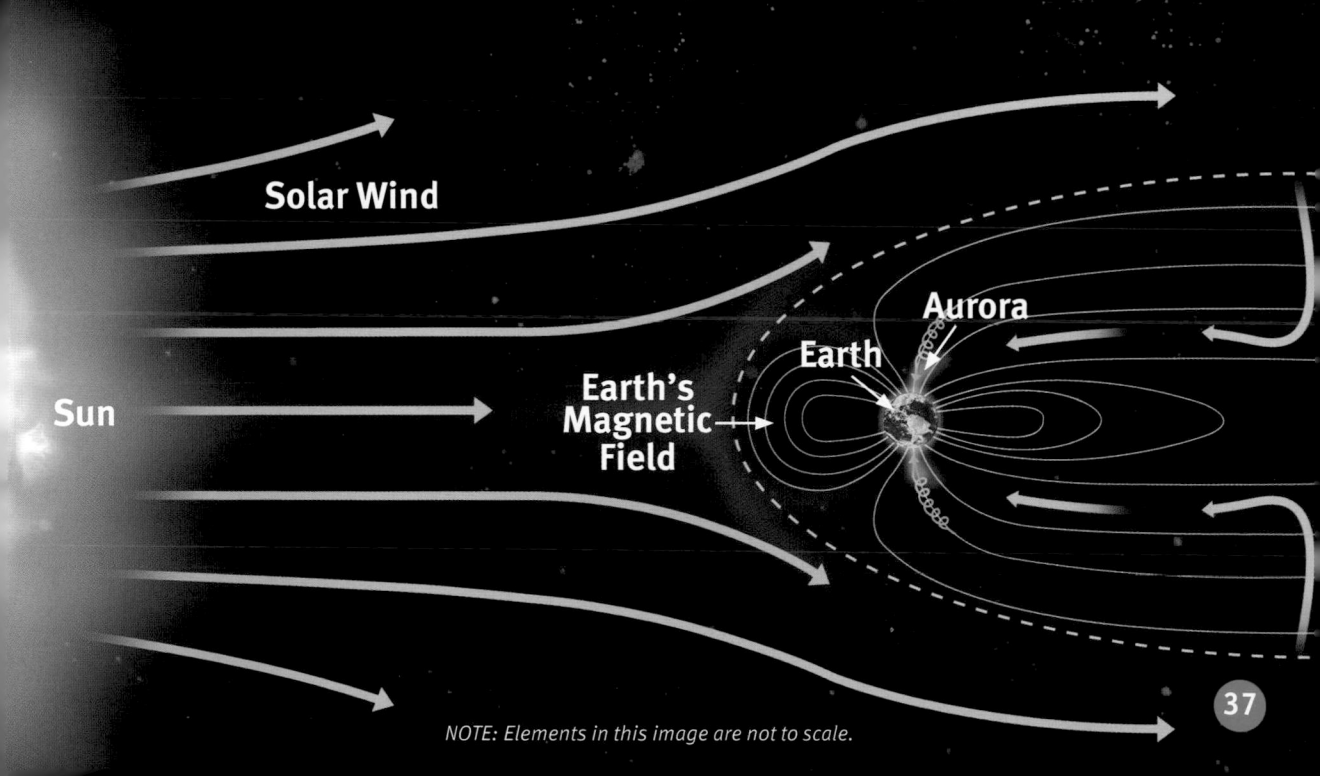

Solar Wind

Sun

Earth's Magnetic Field

Earth

Aurora

NOTE: Elements in this image are not to scale.

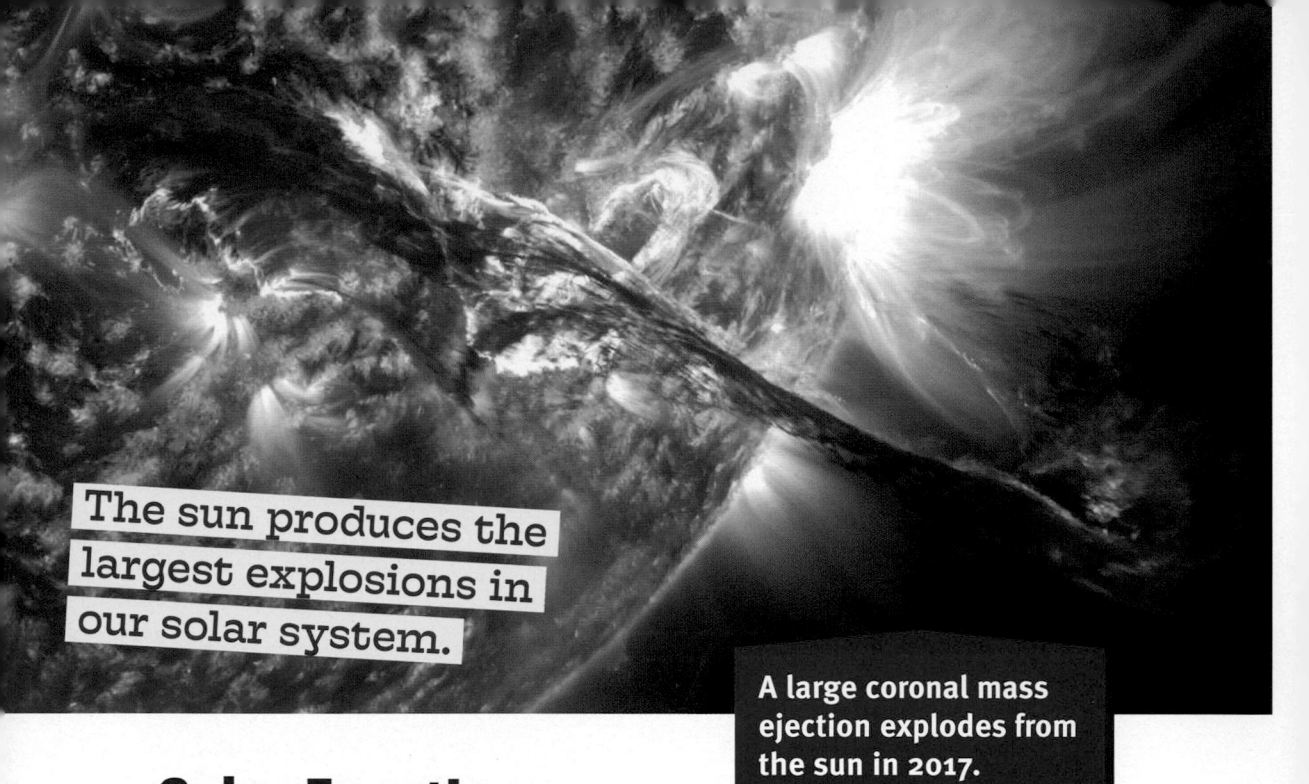

The sun produces the largest explosions in our solar system.

A large coronal mass ejection explodes from the sun in 2017.

Solar Eruptions

As the sun spins, its flowing gases move at different speeds. This can cause its magnetic field lines to get tangled up—and sometimes snap. The pent-up magnetic energy is then released as a solar flare, or bright burst of energy on the sun's surface. Bigger eruptions, called coronal mass ejections, shoot out hot material from the sun. If this material heads toward Earth, it can interfere with satellites, power grids, and communication equipment.

Harnessing the Sun

Besides the occasional outburst, the abundant energy the sun provides is good for life on Earth. We have even found ways to capture that energy using solar panels that convert sunlight into electricity. These devices could someday provide power to people around the globe. Unlike other sources of energy, harnessing sunlight doesn't produce pollution. It is also a **renewable energy** source. This means it will not run out. At least not for billions of years!

Solar panels blanket a mountainside in China.

Tracking Solar Activity

One way scientists monitor the sun's activity is by tracking sunspots that appear on its surface each year. Records of this data go back at least four centuries. The graph at right shows the sunspot counts for a period of 35 years. Study the graph, and then answer the questions that follow.

Numerous sunspots can be seen in this image of the sun.

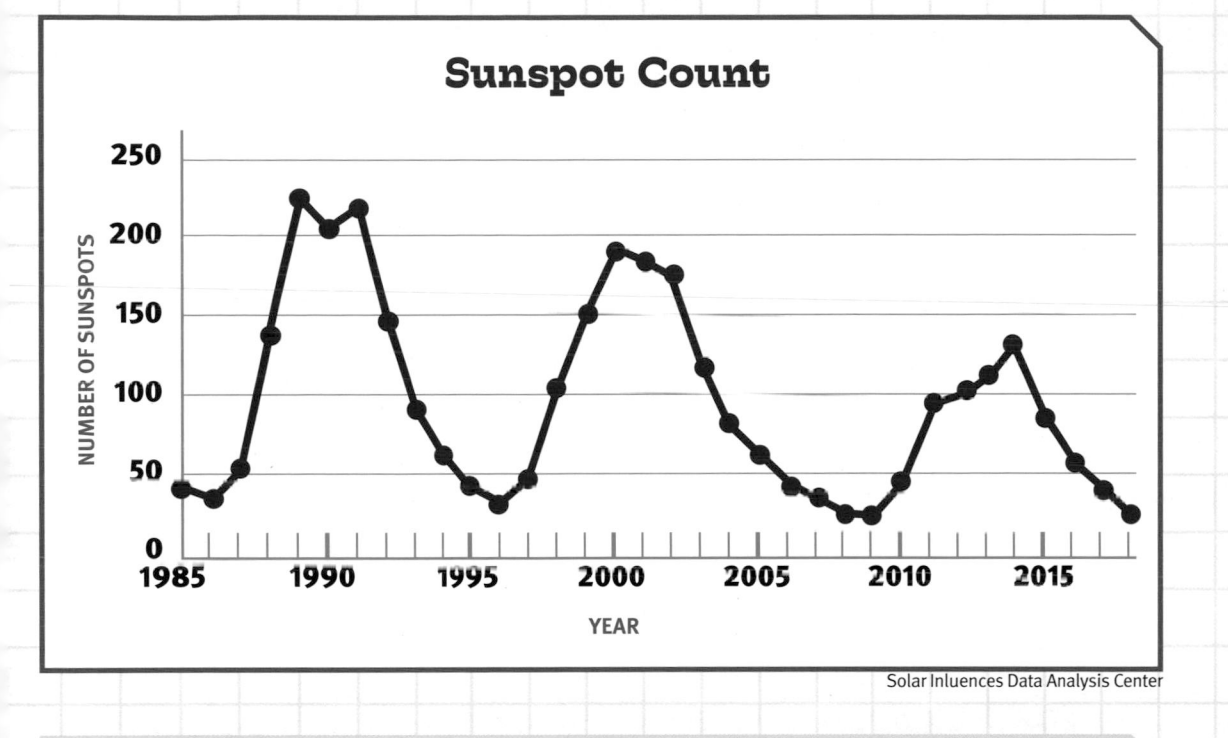

Sunspot Count

NUMBER OF SUNSPOTS / *YEAR*

Solar Inluences Data Analysis Center

Analyze It!

1 About how many sunspots were recorded in 2003?

2 The greatest number of sunspots occurred in which year on the graph?

3 Do you see a trend, or pattern, in the graph? If so, describe it.

4 Sunspot numbers rise and fall about every 11 years. Do you think some of these cycles are more active than others? Explain the reasoning behind your answer.

ANSWERS: 1. About 120. 2. 1989. 3. Yes. The number of sunspots rises and falls periodically. 4. Yes. Some cycles are more active than others because their sunspot counts are higher.

Make Your Own Timekeeper!

Sundials were the earliest devices invented to tell time. They date back thousands of years. Here is how you can make your own!

Materials

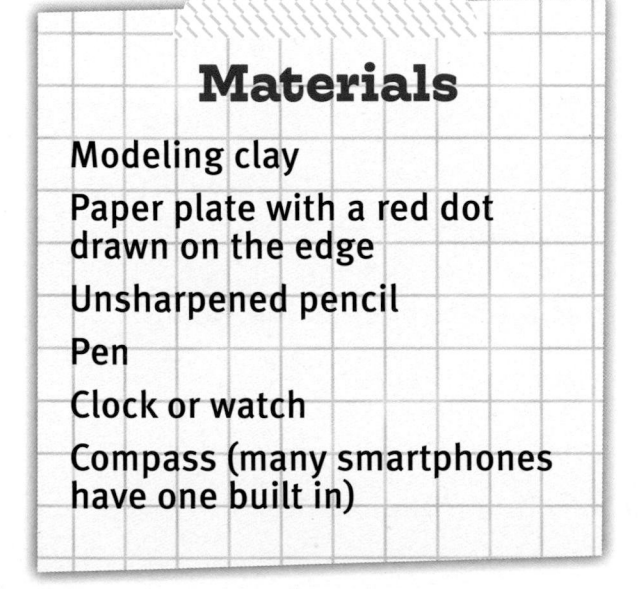

Modeling clay

Paper plate with a red dot drawn on the edge

Unsharpened pencil

Pen

Clock or watch

Compass (many smartphones have one built in)

Directions

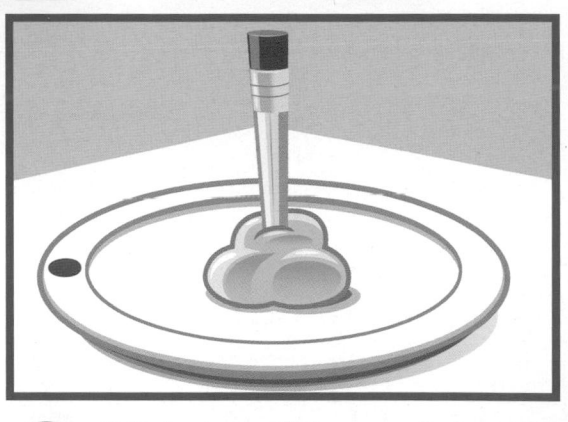

1 Stick a small lump of clay to the center of the plate. Place the pencil into the clay so the pencil stands upright.

2 Early in the morning, take the plate outside. Set it in a flat, sunny spot. Using your compass, find north and make sure the red dot on the plate is pointed in that direction.

3 On the hour, observe the shadow the pencil makes on the plate. Draw a line from the center of the plate to the edge where the shadow falls. Record the hour next to the line.

4 Repeat step three on each hour for as long as there is sunlight outdoors.

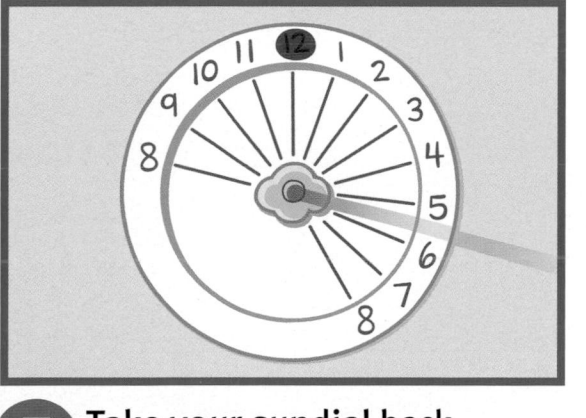

5 Take your sundial back outside the next day. Make sure the dot is pointed north. Observe where the pencil's shadow is falling. Can you use it to tell the time?

Explain It!

Using what you learned in the book, can you explain what happened and why? If you need help, turn back to page 15.

Note: If you are doing this experiment in the Southern Hemisphere, the red dot should be oriented south.

True Statistics

The sun holds this much of the solar system's total mass: 99.8 percent

The number of Earths it would take to equal the sun's mass: 333,000

The sun's diameter: 864,938 miles (1,400,000 km)

Number of Earths you could line up across the diameter of the sun: 109

Average distance from Earth to the sun: 93,000,000 miles (150,000,000 km)

The age of the sun: 4.6 billion years

Temperature at the center of the sun: 27,000,000°F (15,000,000°C)

Temperature at the sun's surface: 10,000°F (5,500°C)

Length of time it takes for the sun's light to reach Earth: 8 minutes

Did you find the truth?

(F) The sun is made up mostly of a gas called helium.

(T) The sun is the largest object in our solar system.

Resources

Other books in this series:

  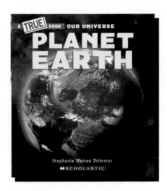

You can also look at:

Betts, Bruce. *Super Cool Space Facts: A Fun, Fact-Filled Space Book for Kids*. Emeryville, California: Rockridge Press, 2019.

McAnulty, Stacy. *Sun! One in a Billion*. New York: Henry Holt and Co., 2018.

Seluk, Nick. *The Sun Is Kind of a Big Deal*. London, U.K.: Orchard Books, 2018.

Glossary

atoms (AT-uhms) the smallest parts of an element

axis (AK-sis) imaginary line through the center of an object

fusion (FYOO-zhuhn) when two or more atoms combine to form a new element

gravity (GRAV-i-tee) force that pulls objects toward one another

mass (mas) the amount of physical matter an object contains

nebula (NEB-yuh-luh) cloud of dust and gas in space

orbit (OR-bit) to travel in a curved path around another object

renewable energy (ri-NOO-uh-buhl EN-ur-jee) power from sources that can never be used up

solar system (SOH-lur SIS-tuhm) all the bodies—including planets, their moons, asteroids, comets, and Kuiper Belt objects—orbiting the sun

star (stahr) shining ball of burning gas held together by its own gravity

wavelengths (WAYV-lengkthz) distances between one crest of a wave to the next

Index

Page numbers in **bold** indicate illustrations.

About the Author

Cody Crane is an award-winning children's writer, specializing in nonfiction. She studied science and environmental reporting at New York University. She always wanted to be a scientist but discovered that writing about science could be just as fun as doing the real thing. She lives in Houston, Texas, with her husband and son.